Off grid Solar Living

Total Solar Conversion for Your Home on a Budget

Outdoor Cooking with Solar

By

Allen Freeman

Published by:

Streets of Dream
Press

Streets of Dream Press
P.O. Box 966
Semmes, Alabama 36575

Cover & Interior designed

By

Jackie Bretford

First Edition

TABLE OF CONTENTS

INTRODUCTION

There is a saying that solar energy might be the last, "gold rush," energy resource for people in the digital age to exploit. Energy produced by sunlight isn't owned by any government. Energy produced by sunlight isn't taxed as a commodity by anyone. The era of renewable energy is dawning, and every homeowner who is regularly taken aback by their home energy bills for electricity should take notice. No, strike that thought, *must* take notice of the potential long-term benefits of home solar energy conversion.

Put simply, the consumption of solar produced energy is the future of residential energy consumption. The practice is already occurring in federally advocated, promoted and financially-subsidized fashions on commercial, residential and government levels. According to statistics compiled by the Energy Information Administration, in 2016 the average American homeowner paid over $112 a month in energy costs for electricity to power the average home measuring 2,500 square foot. Depending on where you live, you are likely to be paying a lot more than the average of $112 a month.

2016 Average Monthly Bill- Residential

(Data from forms EIA-861- schedules 4A-D, EIA-861S and EIA-861U)

State	Number of Customers	Average Monthly Consumption (kWh)	Average Price (cents/kWh)	Average Monthly Bill (Dollar and cents)
New England	6,290,369	616	18.81	115.93
Connecticut	1,486,741	711	20.01	142.19
Maine	699,321	546	15.83	86.48
Massachusetts	2,740,865	599	19.00	113.77
New Hampshire	612,716	604	18.36	110.95
Rhode Island	438,507	586	18.62	109.02
Vermont	312,239	549	17.37	95.31
Middle Atlantic	15,964,597	636	15.68	109.50
New Jersey	3,510,141	691	15.72	108.58
New York	7,118,901	595	17.58	104.58
Pennsylvania	5,335,555	841	13.86	116.67
East North Central	19,937,976	785	13.06	102.55
Illinois	5,231,541	733	12.54	91.83
Indiana	2,821,546	975	11.79	114.96
Michigan	4,311,008	668	15.22	101.64
Ohio	4,911,597	891	12.47	111.15
Wisconsin	2,662,284	683	14.07	96.08
West North Central	9,345,853	917	11.79	108.13
Iowa	1,358,902	864	11.94	103.17
Kansas	1,252,846	899	13.06	117.34
Minnesota	2,376,681	764	12.67	96.79
Missouri	2,751,460	1,041	11.21	116.63
Nebraska	834,038	973	10.84	105.47
North Dakota	377,739	1,046	10.16	106.28
South Dakota	392,187	981	11.47	112.53
South Atlantic	27,213,237	1,107	11.56	127.97
Delaware	419,234	947	13.42	127.03
District of Columbia	259,392	804	12.29	98.79
Florida	9,149,214	1,123	10.98	123.37
Georgia	4,240,421	1,138	11.50	130.87
Maryland	2,288,301	995	14.23	141.53
North Carolina	4,423,532	1,101	11.03	121.44
South Carolina	2,209,783	1,155	12.65	146.09
Virginia	3,362,985	1,120	11.36	127.14
West Virginia	860,375	1,102	11.44	126.10
East South Central	8,248,665	1,198	10.86	130.20
Alabama	2,200,574	1,214	11.99	145.55
Kentucky	1,957,188	1,121	10.49	117.65
Mississippi	1,278,616	1,203	10.47	125.91
Tennessee	2,812,287	1,238	10.41	128.89
West South Central	15,687,117	1,154	10.59	122.17
Arkansas	1,368,867	1,083	9.92	107.44
Louisiana	2,059,699	1,240	9.34	115.79
Oklahoma	1,736,819	1,093	10.20	111.49
Texas	10,521,732	1,156	10.99	127.10
Mountain	9,529,171	848	11.65	98.80
Arizona	2,725,510	1,030	12.15	125.19
Colorado	2,260,068	694	12.07	83.85
Idaho	714,365	953	9.95	94.90

That comes out to about $1,344 a year in annual utility bill payments for energy for the average homeowner. Or, $13,440 in utility bill payments for energy paid out over a decade. Or, for a startlingly clearer perspective of what average homeowners pay for their energy utility bills, that average, monthly American utility bill of $112 for energy consumption amounts to over $26,880 over two decades.

Let that number sink in.

Consider your own monthly utility bills for electricity. Do the math.

Owning a home is a serious and usually long-term financial investment. How long do you plan to live in your home? Unless you have near-term plans to sell your home, most homeowners dwell in their homes and property for years if not decades. Most electricity in the United States is created by burning coal. The era of fossil fuels is ending at a significant escalation.

Now is the time to act. Why would you pay a utility company a fortune in energy bills payments over decades if you didn't have to?

Why not invest in solar energy produced technologies so you can create the energy you need to power your home? Your home, which is your multi-decade investment for comfortable habitation and living.

Investing in home solar energy conversion is an appreciable investment. However, there are numerous government-sponsored ways to offset costs. Solar energy is an investment, like your home itself, which will help you power your home on your terms. Home solar energy conversion is an investment

MY RESEARCH

As a homeowner I was skeptical about changing the way I powered my home. I was used to paying for electricity, like any other homeowner. The idea of converting my home to be completely solar energy generating seemed too fringe or fad-friendly. A bit too, "new age," for my taste.

I think as human beings, it is hard to let go of the usual way of doing things. I realized, and accepted, that the usual way of doing things in the public and private energy sector was coming to an end.

Owning a home is a business investment and paying for energy, like electricity, can be expensive. It's even worse considering how much energy I paid for that was wasted. I was tired of worrying about how to pay my home energy bills. I also developed a homeowner's conviction to only pay for the exact amount of energy that I used and/or needed, and not a watt more than that.

My research into home solar energy conversion began in earnest when I accepted the fact that the renewable energy industry will only continue to grow. Now is the best time in decades to affordably buy a residential solar energy setup. Converting your home energy setup to solar energy is expensive, there is no question about it. Still, there are many readily available, solar energy friendly government tax incentives available. I saved 30 percent on my initial home solar energy conversion purchase costs. The United States government wants residential and commercial building owners to adapt to solar energy and will give them every financial incentive to do so.

The 5,000-kilowatt solar panel array installed on my roof cost me about $17,000. My home is about 2,500 square feet, which is the national average. How much you will actually pay for your home solar energy conversion will vary state-to-state and how much sun exposure your region experiences. So, think of it this way – when you take out a mortgage on your home that is a long-term investment as well. Initial home solar energy conversion costs are comparable to the purchase of a brand-new economy grade vehicle.

My home solar energy setup will pay for itself in a few years. Most importantly, after I sought out every solar-friendly tax incentive that applied to me, my initial purchase price of $17,000 was knocked down to $11,000. I am still hooked up to the utility grid, but I only draw power from it at night or incidents of inclement weather. That is unless I have enough energy stored in my reserve battery. Moreover, through a process called, "net metering," any excess electricity that I generate is sold to the utility company. Yes, that's right – the power company pays me! However, this is via non-monetary credit, as opposed to the popular fiction that utility companies pay for excess electricity generated by residential solar energy systems.

Still, free energy is free energy.

So why am I saying of all this and throwing figures at you? It is not my intent to make this process sound easier than it is. Solar energy home conversion is quite an investment, and it will take years, if not decades, for it to pay for itself. On the average, it could take anywhere from five to twenty years for a solar energy system to pay for itself, based on your energy consumptions needs, type of setup and where you live.

Yet, the initial setup costs are worth it, especially if you take advantage of all available and applicable, solar energy friendly investment tax credit incentives. Like me, you will be taking charge of your energy needs, instead of paying some utility company for it. Consider this – based on current rates; you stand to save anywhere between $40,000 to $90,000 on energy costs over a period of two decades with home solar energy conversion. Of course, this would all depend on where you live, the size of your solar panel roof array, how much energy you averagely use, and the depth of tax incentive breaks you take advantage of, among other things.

Scared yet? Yes, this is a momentous decision and undertaking to consider.

Change is hard. It is hard to look at your home and then decide to make big changes. What you should fear is not embracing change and not accepting the fact that the reign of fossil fuels is coming to an end. In many ways, it already has. As a homeowner, you should adopt as well and plan accordingly.

You can harness energy on your own terms via your own investment. Why not?

Oil and gas prices will continue to fluctuate. Most electricity in the United States is generated by burning coal. Most electrical power grid infrastructures in the United States are outmoded, crumbling and will have to be overhauled in the future.

This may sound like bits of information from a cable news financial network, but they are true, and they will affect anyone who owns a home. Like you and me.

The ways in which people use energy on this planet is changing. This will affect homeowners. It will affect how homeowners power their homes and how much they pay for power. The energy business is based on the supply and demand model. Fossil fuel use will not end overnight, but it will continue its steady, and precipitous, decline throughout the next half-century as supplies steadily dwindle. Prices will fluctuate, and spike, as per the tenets of supply and demand during this time.

That reality will tremendously affect everyone who hasn't adapted to renewable energy, like solar.

As a homeowner myself, I think you should get ahead of the curve because the age of renewable energy is upon us. Take charge of your energy needs. No, it won't be easy, but it will be worth it. Your home is an investment, and sometimes investments need to be upgraded. I will present the facts and figure to you in this book, and you will see for yourself the very calculable cost benefits of home solar energy conversion.

Sunlight is not taxable and can't be commoditized by any government. You can harness as much of it as you can for your own energy needs. As a homeowner, you live in the moment of time where converting to solar energy consumption is as affordable as it ever has been. Solar friendly financial and tax credit incentives are readily available for homeowners for the taking, for the time being. Solar powered energy is yours for the taking. You can, and you should.

Solar energy will indeed become the closest thing to a "gold rush," resource in the digital age. It will probably be the last one. Comparably, for most people, home ownership can sometimes be considered the last great investment of a lifetime. Embrace home solar energy conversion and optimize your home ownership investment, and energy usage, for the long run.

THE FINANCIAL INCENTIVES OF HOME SOLAR ENERGY

You may have heard of the saying, "burning the candle at both ends," or, "burning the midnight oil." Both sayings are very apt when it comes to the rapid rate at which humans are using up available global reserves of fossil fuels. In 50 or 100 years, there just won't be enough oil on a global scale to conduct business, or affordably heat your home, as it was done in the past.

As a homeowner myself, I knew the time had come to seriously consider the benefits of, "burning the light," so to speak. According to the United States Department of Energy, on the average, a homeowner who converts to solar energy stands to meet over 40% of their energy needs via solar. Depending on your own solar energy home setup, such an amount can be more or less.

Solar energy represents a future liberation of homeowners from the commoditization of fossil fuel energies. The very act of homeowners buying energy should become an outmoded utility expense of the past. We live in an era where such a concept is possible. There has never been a better time in history to convert your home to make use of solar energy in an extremely budget-friendly manner. If you have been considering adapting your home energy consumption needs for the use of solar energy, then right now is the time to do it.

The best place for you to start investigating the cost benefits of home solar energy conversion is through the use of solar-friendly government tax credits and incentives. You can, and should, take advantage of solar-friendly tax incentives on the local, municipal and federal level. They are widely available. The United States government wants to give you every financial break possible to adapt and convert to home solar energy conversion.

This exact moment in time is the prime moment in proprietary energy and utility history to take advantage of solar energy friendly government tax incentives. Currently, there is an over-abundance of available government tax incentive and rebates to help you recoup up to 30% of your initial solar energy conversion set-up costs.

If you plan to commence building a DIY solar energy, off-grid home construction project, there has never been a more budget-affordable time to buy home solar energy panels, equipment and setup gear. You may be able to save more if you look for every tax incentive that applies to you on the local, municipal and federal level. I am speaking from experience when I tell you; there are just too many ways to save money on home solar energy conversion right now.

The best place to start on this issue is the investment tax credit initiatives for residential, commercial and utility conversion. The United States government wants you to convert to solar energy for as much of your home energy consumption needs as possible. However, unless you run a business or utility company, homeowners have only until 2021 to take advantage of numerous solar energy tax credit initiatives.

I am sure that the mere mention of, "tax credits," may be enough to limit attention spans or cause eyes to glaze over. I assure you, it would be a mistake to do so. This is something worth knowing about if you own a home and pay utility companies for energy use.

Introduced into the American tax code in the late 1970's, investment tax credits have become highly relevant in the current era of renewable energy market ascendancy. Quite simply, solar energy advocating investment tax credits are your financial threshold to budget-friendly, home solar energy conversion cost savings.

INVESTMENT TAX CREDITS

Written into the American tax code in 1978, renewable energy advocating investment tax credits were implemented to encourage American businesses, residents and power companies to explore actionable ways to use less fossil fuel based energies like oil, coal, and gas. Investment tax credits provisions, which are provided by Internal Revenue Code, Section 48, were created in the 1970s during the energy crises and shortages of the period. **(1)**

They are designed to offer recoupable, financial relief for the initial purchase costs, as well as materials and construction costs, related to the investment of renewable energy technologies, like solar. ITCs were also legislatively envisioned as a way to organically expand the widespread acceptance and implementation of renewable energy technologies and initiatives.

OK, so what does that mean to you as a homeowner trying to save on energy expenses? The United States government wants to give you every chance to save as much money as possible on personal investments related to converting your home for solar energy use. Global, government, industrial and residential expansion into solar energy is occurring at a quick pace. The solar energy investment tax credits were scheduled to end in 2016 but were congressionally extended until the year 2021.

For the purposes of estimation and providing examples in this book, I will mainly be using estimations referencing an average American home of 2,500 square feet that is utilizing 5,000 kilowatts of solar energy. The national average cost per watt for solar panel installation in 2017 was $3.16 per watt. **(2)**

That comes to $15,800 for our solar energy home conversion estimate. If we subtract 30% of $15,800, we get $4,740. After taking advantage of a solar energy investment tax credit rate of 30%, the estimated initial cost can be knocked down to $11,060.

Of course, this is just an estimation for the purposes of example. How much you may pay, or save, will depend on where you live, how much sunlight your region receives and a few other factors that we will discuss in later chapters.

Barring any further action taken by the Congress of the United States, you only have until 2021 to take advantage of the 30% solar energy investment tax credit. In 2020, the ITC will lower to 26%. In 2021, the ITC will lower still to 22%. After December 21, 2021, the ITC for homeowners will be phased out completely. In 2022, only utility companies and commercial businesses will enjoy a permanent 10% investment tax credit from that point onwards.

1 Sherlock, Molly F. "The Energy Credit: An Investment Tax Credit for Renewable Energy." *Tax Equity Times*. October 7, 2016. Accessed November 30, 2017, https://www.taxequitytimes.com/wp-content/uploads/sites/15/2016/10/CRS-2016-ITC-Report.pdf

2 EnergySage http://news.energysage.com/how-much-does-the-average-solar-panel-installation-cost-in-the-u-s/

When I said that this was the best time in history to get in on solar energy, I really meant it. If you are really considering getting in on solar energy, then it is best to get in now, "while the gettin' is good," so to speak.

Figure 1.

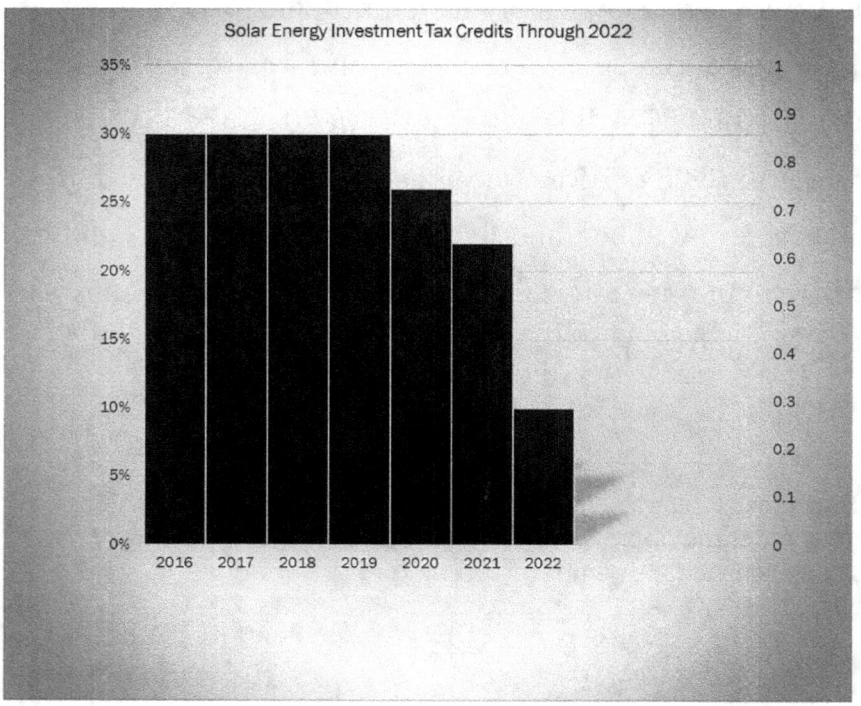

(Please note that the 2022 initiating, and permanent, ITC rate of 10% will only be applicable for use by commercial entities and utility companies. All applicable ITC rates for residents will fall to 0% after the year 2021.)

Believe me, I know, converting your home for solar energy usage will be a time-consuming endeavor. You will save over the long run, but it is costly up front. There are many things to consider.

However, 30% off your initial start-up costs in relation to solar energy home conversion should be a slam dunk decision. Use this e-book as a launching pad for further research, take some time to think about it. But don't waste too much time.

To further accentuate this point, I want to go through some real-world examples of the type of potential savings that you can expect when you convert to solar energy.

SOLAR ENERGY CONVERSION SAVINGS CALCULATIONS

Using some basic location calculation searches on the internet, you can roughly estimate how much you will save on home solar energy conversion in years or even decades. (3)

I want you to use this e-book as a launching pad for your own research. You should definitely see what you are getting into before you convert your home for solar energy consumption.

EnergySage is a solar energy advocacy company that educates the public about the benefits of solar energy. The company, which is funded in part by the United States Department of Energy, also strives to educate the public about the need to convert to the widespread use of renewable energy sources, like solar, sooner than later.

For your consideration, I will now detail real-world solar energy conversion contractor estimate examples, based on actual customer estimates based on EnergySage calculations, for five major American cities. I would like for you to see for yourself the potential real-dollar amounts you could save on solar energy conversion investment costs. Also, while some areas of the country receive more sunlight exposure than others, solar energy is a sound investment, wherever you live.

According to estimates as calculated by EnergySage and the Department of Energy, a homeowner with 2,500 square feet of home space and converts to a 5,000-kilowatt solar panel system can save over $100 a month over current electricity costs.(4)

Of course, how much you currently pay for electricity depends on where you live, how much your utility charges for electricity per month in kilowatt hours and how much electricity you consume.

1. New York, New York.

This estimate is based on a customer with a monthly $192 electricity bill. The initial investment cost for solar energy conversion would cost $30,832. The solar energy conversion would pay back for itself in 5-years. This customer would potentially save $34,098 over a 20-year period.

Now consider the 30% solar energy investment tax credit. 30% of $30,832 comes out to about $9,249. That means that the initial startup costs would be scaled down to about $21,583. A $192-a-month electricity bill comes out to about $2,304 annually. Over a 20-year period, that annual $2,304 will amount to $46,080.

3 EnergySage https://www.energysage.com/solar/calculator/

4 EnergySage http://news.energysage.com/whats-financial-value-solar-residential-customers-part-1-rank ng-americas-50-largest-cities/

2. Tampa, Florida.

This estimate is based on a customer with a monthly $90 electric bill. The initial investment costs for home solar energy conversion would amount to $13,642. The solar energy conversion system for this customer would potentially pay back for itself within 8 and half years. This could potentially amount to about $15,188 in energy cost savings over a 20-year period.

Now, let's consider the impact of a 30% off solar energy invest tax credit. When we subtract $30 from $13,642, we get $4092. That means we have knocked down the initial conversion cost here down to $9,550.

A $90-a-month electricity bill comes out to $1,080 annually. That amounts to $21,600 over a 20-year period.

3. Tucson, Arizona.

This is a real-life solar energy conversion estimate for a customer with a $170-a-month electricity bill. The initial cost of conversion would cost $26,676. This solar energy conversion could potentially pay back for itself within 7-and-a-half years. This customer could potentially save $38,399 over a 20-year period on energy costs.

Let's factor in the 30% solar energy investment tax credit. 30% of $26,676 is $8,002. That means we have brought down the initial conversion cost down to $18,674.

A $170-a-month electricity bill will cost $2,040 annually. That amounts to $40,800 over a 20-year period.

4. San Francisco, California.

This real-life solar energy conversion customer estimate is based on a customer with a $175-a-month electricity bill. The initial cost of solar energy conversion would cost $19,428. The estimation is that this conversion would pay back for itself within a 6-year period. Over a 20-year period, this customer will potential save $41,059 on energy costs after converting to solar.

Now for the 30% investment tax credit deduction. When we deduct 30% from $19,428, we get $5828. That means that the initial cost for this customer's home solar energy conversion has now been knocked down to $13,600.

A $175-a-month electricity bill costs $2,100 annually. That same electric bill amounts to $42,000 over 20 years.

5. Houston, Texas.

As with the previous examples, this is a real-life customer who made solar energy calculations with EnergySage's online savings calculator. This customer has a $108-a-month electricity bill. The initial home solar energy conversion cost was estimated to be $15,200. This solar energy conversion estimate could potentially pay back for itself within 8 years. This customer could save over $15,948 over a 20-year period on energy costs by converting to solar.

By taking advantage of the 30% solar energy investment tax credit, about $4,560 can be deducted from initial startup costs. This is means that this customer could end up only paying $10,640 after investment tax credit rebates.

A $108-a-month electricity bill comes out to $1,296 annually. That amounts to $25,920 in electricity bills over a 20-year period.

Of course, these are just projected customer solar installation estimates that are based on online calculations. Your savings potential will differ based on variables like where you live, how much you pay for utility bills and the projected cost and size of any potential solar energy system you plan to install.

This graph, compiled by the Solar Energy Industries Association, details the top-ten states in 2017 that utilized the most out of renewable solar energy.

Figure 2.

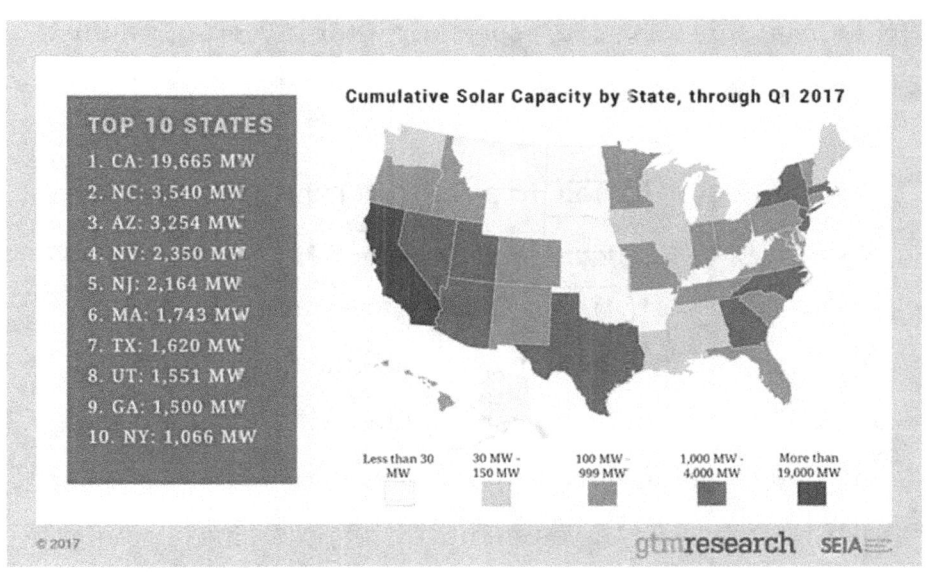

(Source: SEIA)

BUYING VS. LEASING

I highly recommend that you look into all available financing options that will allow you to outright buy your residential solar panel system. The only way to take full advantage of solar energy investment tax credits is to outright own your solar panel system.

If you lease your solar panel system, then the leasing company will own the system and be eligible for all applicable commercial investment tax credits related to solar energy. Why would such a situation make any financial sense for you, especially after investing so much financially to convert?

You may be able to pay for your home solar energy conversion costs with a home equity loan, assistance from a credit union authority or some other financial lending authority. What's more, without taking advantage of all available, and applicable, investment tax credits, you may end up losing more money than you save on a home solar energy conversion investment. If you opt to lease your home solar energy conversion, your setup may not pay for itself in an appreciable amount of time.

Also, take great consideration in considering how long you will actually live in your home. Home solar energy conversion is an investment that will pay for itself in years or decades. So, as long as you don't plan on selling your home in next 15-20 years, this is a sound investment. This way you can make sure your financial investment will be worth it in the long run.

If you ever plan to sell your solar energy converted home, you will have to look into the legal transference of your contract with the utility company to the new buyer. If your system is leased, there will probably be bureaucratic processes, and paperwork, to endure to switchover ownership. You may be required to provide prior written notice of contract transferal of the leased solar energy system or even endure drawing up new contractual transference paperwork.

Such a situation will just become an added headache if you ever sell a solar power energy consumption converted house with a leased system.

Take it from me, your home and how you power it is a long-term investment. If you decide to convert to solar energy produced power, I highly suggest you buy and own your system outright.

Figure 3.

(**Source:** Pixabay)

NET METERING

Net metering is a federally mandated energy-crediting process designed to deal with the over-production of solar power generated energy.

As of the year 2017, and according to EnergySage, over 40+ American states, the District of Columbia and several U.S. territories, all mandate net metering protocols for residential homeowners who have converted to solar energy use. **(5)**

The process of net metering was designed to accommodate residential, solar energy converted homeowners who are still connected to a utility grid to make the most of the over-production of energy, as well as periods of energy shortages.

The one basic, and obvious, limitation of solar-generated energy is that it does not always generate electricity. Solar panel arrays do not produce electricity during the night or during incidents of cloudy or inclement weather. For example, snow-covered solar panels do not generate energy. Further, when you convert your home for the use of solar energy, you may opt to design your system to use only as much energy as you previously did when you paid for electricity. There may be times when your solar energy system generates more energy than you need or will use, at a given time.

With net metering, any excess energy that your solar panel system generates will automatically be fed back into the utility system grid. Your meter will actually run in reverse according to how much excess energy your solar panel system feeds back into the grid.

Under net metering, you will be automatically credited for the amount of energy you feed back into the grid. In such a situation, your meter running in reverse will help keep track of how much excess energy your solar energy system is putting back into the utility grid. Then, at night, in instances of inclement weather, or periods when you use more energy than you generate, this credited, excess energy will be fed back into your system by the utility company.

It's best to think of net metering as a utility grid-wide efficiency system to make the most of the energy generated by solar energy. According to estimates compiled by the Solar Energy Industries Association, as little as 20%

to as much as 40% of solar power generated energy is ever directed onto a utility grid via net metering. **(6)**

If by chance your home uses more energy than you feed into the utility grid system, then you will be billed by the utility company.

5 EnergySage https://www.energysage.com/solar/101/net-metering-for-home-solar-panels/

6 SEIA https://www.seia.org/initiatives/net-metering

In this situation as long you your home isn't drawing upon electricity from the grid, you will be billed via credit for your, "net," usage of credited energy. When you convert your home for the use and consumption of solar energy, you want to take several things into account.

Before your conversion, you may take into account how much energy you consumed and how much you want to consume. Even with adjusted consumption calculations for your solar energy home conversion setup, you may end up creating more energy than you need. Plus, there may be periods where you don't have enough accrued net credits to draw energy from.

Figure 3.

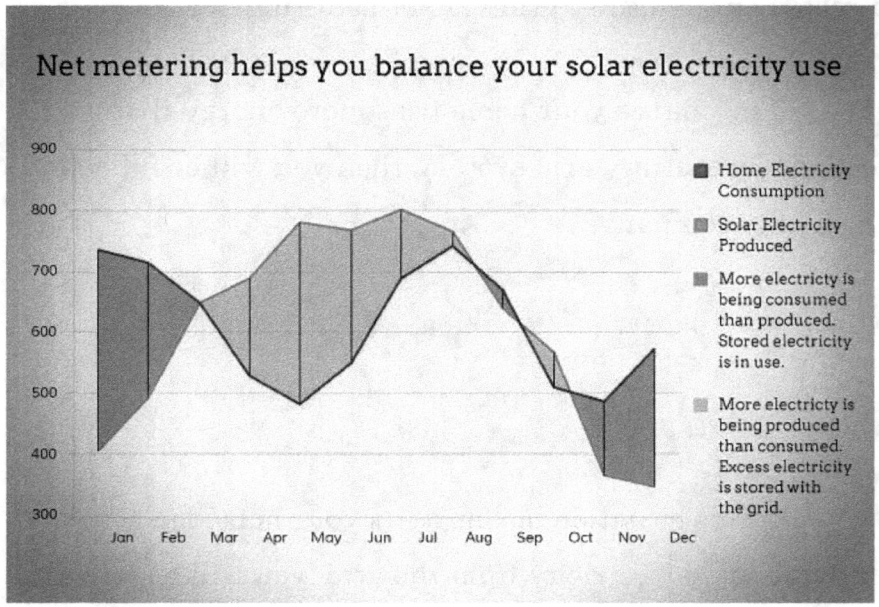

(**Source:** EnergySage)

According to EnergySage, and my own experiences with net metering, there is absolutely no such thing as the utility company cutting you a check or paying you cash money for the excess energy that your solar power system generates.

Net metering is a wholly non-monetary method of a utility company crediting a homeowner for the amount of energy that is fed back into a utility company's energy grid. As of yet, renewable solar energy has yet to commoditized in such a way by anyone, authority or government.

To stress this again, if you take advantage of net metering, you will only receive utility energy credits for any excess energy that your solar energy system produces. It is basically a self-crediting account system to ensure that you are never without power when you need it and can store any excess power you generate.

As much as I wish it were true, the utility company will not cut you a check for hard cash for your excess energy via net metering.

THE BASICS OF A HOME ENERGY GRID

As a homeowner mulling home solar energy conversion, I want you to make the most out of, "burning the light," so to speak. I think it may help to generally break down exactly how solar energy would work in a home energy grid should you decide to convert your home.

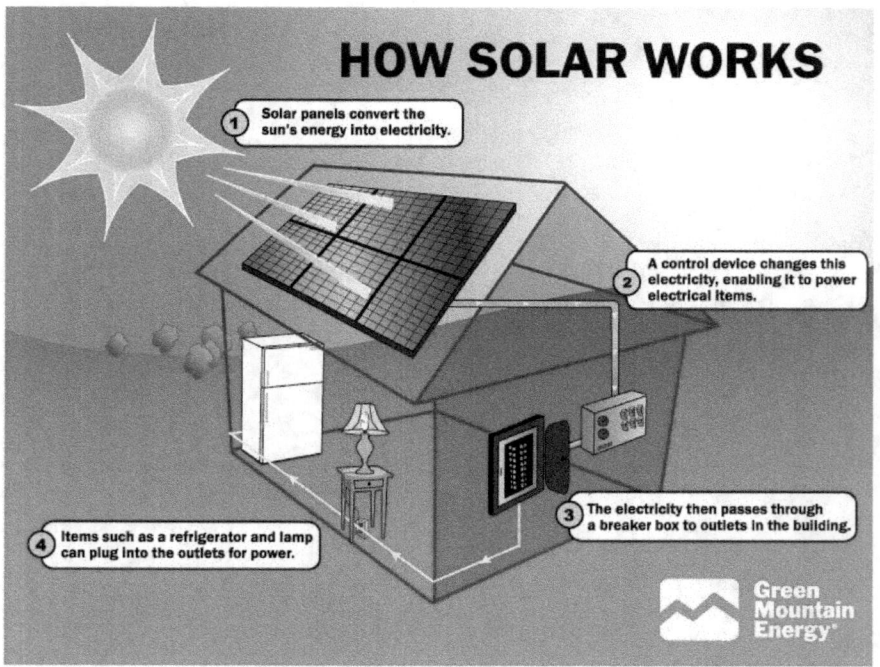

(Credit: progressivce-charlestown.com)

Your basic home solar energy grid will comprise of wiring connections between power lines owned by the utility company (as long as you are connected to the grid), a connection to the grid, solar panel arrays, an inverter, and meter to record consumption and for the purposes of net metering.

I highly suggest that you thoroughly research and contract a recommended, licensed, insured, professional solar panel system installer and licensed electrician to install your solar power system. Converting your home to consume converted solar energy is no easy task and involves a lot more technical work than just applying solar panels onto your roof. After all, this is a significant investment that is worth doing right the first time.

THE BASICS OF SOLAR ENERGY

A solar panel system, like the kind that could potentially end up on your rooftop, is more accurately known as a "photovoltaic system." So, what does, "photovoltaic," mean?

It is a system by which sunlight, or photons, are captured in a solar panel and then converted into electrical energy. Your solar panel may be made from some variation of a crystalline silicon wafer. Quite simply, solar panels are a semi-conducting material with interior electrical circuits.

As sunlight strikes the solar panel, electrons within the semiconducting material begin to move around. These excited solar-excited electrons create an electrical charge that can then be directed through electrical circuits in the panel and collected as energy.

Figure 4.

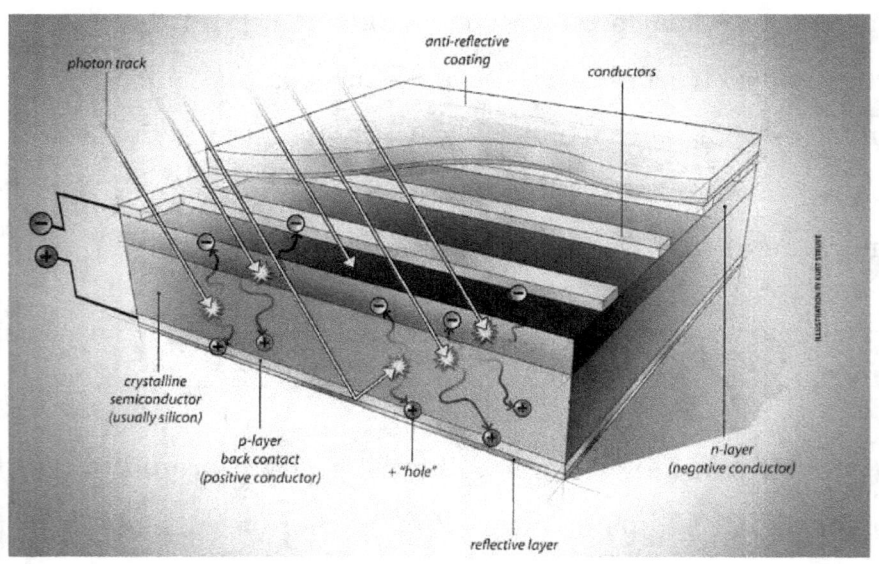

(Credit: SEIA)

Multiple solar cells can comprise a solar panel, while a grouping of solar panels can comprise a module or array. How many solar panels you will require on your roof or property will depend on where you live, how much energy you use, how much energy you want to use and how much sunlight your region generally experiences.

OK, so the solar panels are on your roof, receiving sunlight which is exciting electrons within the panel, which in turn is generating electricity. Well, this solar energy generated electricity is not usable home energy use needs. Not just yet. Photovoltaic, or solar energy, generating systems create D/C, or direct current electricity. Almost all modern appliances and the wiring in your home is designed to accommodate A/C, or alternating currents of electricity.

After your solar panel array or module is installed on your roof, you will have to have an inverter installed in your house or property. An inverter is a device that switches the D/C current of electricity that is generated by your solar panel system into an A/C current.

Wiring will have to expertly, professionally and safely connected from nearby power lines, (if you opt to be grid connected) your solar panels and to your inverter. Without an inverter device, you will not be able to make use of the D/C current electricity generated by your solar panels. That is unless you use appliances or have a house wired to accommodate D/C electrical currents.

As long as you are connected to a utility grid, and live in a state that mandates net metering, your solar panel array will be connected to a meter. This will allow you to track how much energy you consume. Also, via net metering, you can be credited for any excess energy that you feed back into the system

Figure 5.

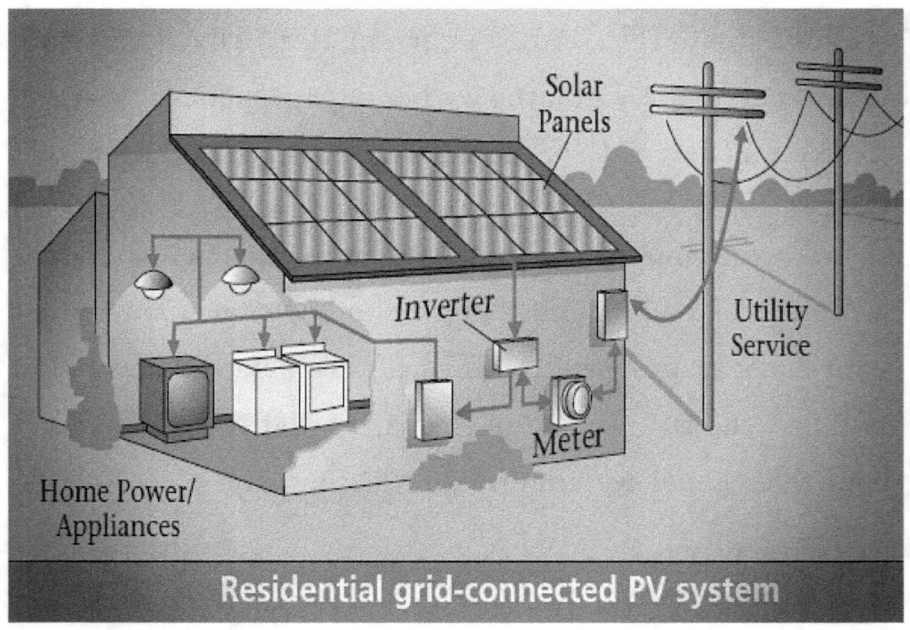

Residential grid-connected PV system

(**Credit:** United States Department of Energy Consumer Guide)

SELECTING A CONTRACTOR

While I will touch upon this subject a bit more in the subsequent DIY chapter, installing solar panel arrays on your roof or property may be a job that is best left to experienced and expert solar installation contractors. Installing solar energy generating panels is not an easy task. It is not as simple as affixing solar panels to your roof and then generating energy. If you opt to be utility-grid connected, there is a lengthy, bureaucratic process involved in connecting your solar energy converted home to the grid.

There are contracts, permits and fees and the local bureaucratic process of connecting your solar panel energy system to the grid to deal with. Most importantly, there is the process of selecting a contractor.

6 EnergySage https://www.energysage.com/solar/buyers-guide/how-to-choose-a-solar-installer/

I went through this process, and I highly recommend that you take your time choosing a competent, experienced, licensed and recommended solar panel installer. Even with the benefit of solar energy investment tax credits, the conversion of your home energy use into renewable solar represents a significant, long-term investment. Take the time, compare contractors and estimates and do it right.

First of all, review and compare the rates of multiple installers. Try to compare at least four licensed, certified, recommended and experienced solar energy system contractors. According to EnergySage estimates, you could save up to 25% on contractor estimates. It always pays to shop around when considering such an investment, whether it be an automobile purchase, mortgage loan or a home solar energy conversion project

The contractor or contracting company that you hire to perform your home solar energy conversion should have extensive experience in performing the procedure. They should come highly recommended by friends, colleagues, family or neighbors. I don't recommend that you randomly choose a contractor for such a project. This is about powering your home for decades to come so choose carefully on your own terms as a homeowner.

The solar energy system contractor that you hire should have a professional reputation that eminently precedes your initial consultation. Look up the history of the contractor online. Check with your local, municipal or federal consumer protection authorities and see if the contractor you will potentially hire has a history. Believe me, I did this myself. Check them out.

Your solar energy system contractor should be able to show you real-life, evidential examples of previous solar energy installations for previous clients. Your solar energy system contractor should also possess several kinds of commercial licenses to perform solar energy system installation. Your potential solar energy system contractor should possess an electrician's license, a general contracting license, and a home improvement contractor's license, to name a few. They should be commercially licensed to perform solar energy installations in the state that you reside in.

Inquire if your solar energy system contractor has any commercial and/or professional certifications. Are they certified to perform solar energy installations on residential properties by the North American Board of Certified Energy Practitioners? The NABCEP also oversees the issuance of the Solar PV Installation Professional Certification. When you are confident that your contractor possesses some kind of professional certification for their line of work, you may be more at ease knowing that they know what they are doing.

Your home, and solar energy conversion project are momentous investments, as I continually remind you.

I know, my personal solar energy installer was NABCEP certified. If your potential solar energy system installer does not possess certification, they should at the very least own an experienced, reputable and provable track record of their skills.

Make absolutely certain that your contractor is adequately insured and bonded to perform their work. Your contractor should at the very least possess a general form of commercial liability insurance.

Talk through the entire process with your contractor before you sign any contract or pay any permit fees. Read any contract or agreements thoroughly before you sign it. Make sure there is a multi-decade warranty on the solar energy system and panels that you have installed on your home. Solar panels are usually warranted for a decade or more.

However, you should make absolutely certain that you specify warranty length for your solar energy system and all related equipment in writing for a decade or two.

This is a long-term investment, considering how long you may end up dwelling in your home.

Remember that there may be a period of weeks or months after the signing of contracts and before actual work on your solar energy home conversion system begins. Converting your home to accommodate the use of solar energy, reconfiguring your home's connection to a utility power grid, sourcing and installing equipment and finalizing bureaucratic paperwork, permits and contracts take time.

This link is for a 2-minute YouTube video produced by the United States Department of Energy. It is called, "Energy 101: Solar PV." It will give you a basic, introduction to how solar power works. It is also a great starting point for you to begin learning more about the subject.

United States Department of Energy – "Energy 101: Solar PV"

https://www.youtube.com/watch?v=0elhIcPVtKE&index=19&list=PLACD8E92715335CB2

DIY HOME SOLAR ENERGY SETUP

Solar energy power generation allows homeowners the ability to live a self-sustaining lifestyle. I have been talking about home solar energy conversion as related to the average residential homeowner who is still connected to a utility grid for net metering purposes. Now I will touch on installing your solar energy system yourself in a DIY manner.

Whether you opt to be grid connected is up to you. Although I highly recommend that you do connect your system to a utility grid and have your DIY installation project professionally inspected, which is a requirement of law anyway.

I highly recommend that if you opt to self-install a DIY home solar power setup that you have or attain, some experience or education in home improvement remodeling, repair, solar energy system installation and/or electrical work training. Or that you retain the advice, counsel, and suggestions of a licensed, professional and experienced electrician or solar energy system installer.

Solar energy system installation requires the applied use of mathematics and the need to be mechanically inclined to work with complex electrical systems. Self-created problems and mistakes during the installation process will be costly and dangerous.

You want to be as safe as possible and make sure that any solar energy system that you install is installed as expertly and professionally as possible. Hazards, defects, and system flaws have to be recognizable when apparent, as electrocution and fire are constant threats, especially against a potential off-grid system. Your home, whether connected to a utility grid or off-grid, represents a significant financial investment. Your home is a symbol of your future.

Unless you're an experienced electrician or home improvement expert, installing a solar panel system on your roof or property can be a dicey proposition. You should really be an expert electrician or solar panel system authority if you are going to try to build your own solar panels. (Which I do not recommend.) You will probably have to invest in a battery for solar energy storage.

Your DIY solar energy system equipment may not have the warranty protections afforded a solar energy system as installed by a professional contractor. Vigilant, operational maintenance of your solar energy system must be regular, thorough and exacting.

You will need to invest in a battery if you opt not to be grid connected. With a DIY solar energy system installation project, you will have several equipment parts that will have to work in calibrated unison with each other day in and day out. You should have a licensed, professional contractor visit your property, survey it and perform an electrical inspection of your system.

You are legally required to do so in your city, county or municipality, depending on where you live. Especially if you apply for solar friendly investment tax credits, which you can still do, even with an off-grid solar energy system project.

As well as being eligible for 30% deductions on your initial setup costs, investment tax credits are also designed to ensure that homeowners endeavor to have their solar energy systems professionally and safely installed.

I will describe to you how to set up a rudimentary, DIY solar energy system as best as I can, but it would behoove you as a homeowner to take every possible precaution to make sure you are installing your solar energy system as properly as possible.

SOLAR PANEL HOME KIT

You can purchase various solar panel home kits online or in-home supplies departments for as low as $5,000 or as much as $20,000, if not higher. When you take into account the 30% deduction potential of solar-friendly investment tax credits, you will stand to save a lot of money on initial setup costs.

On the other hand, with this DIY setup, you are installing a solar energy power system via a home installation project for long-term use. There are a lot of variables to consider.

(Credit: Pixabay)

- **Will you employ the use of a diesel generator?** If you are not connected to a grid, you may need a backup generator in cases of power outages. You will also have to invest a lot of time management in the vigilant maintenance of a diesel generator. Generators breakdown and require repair from time to time.

- **Fuel.** If you opt to maintain a diesel generator, you will have to calculate how much fuel you will need on hand in reserve for backup and emergency purposes. This will get expensive very quickly if you opt not to be grid connected.

- **Solar energy batteries.** Solar energy batteries can wind up costing you hundreds or thousands of dollars per device. Solar power system battery charging technology and efficiency is still being perfected as well. They require a lot of operational maintenance and repair vigilance. You must consider and calculate your daily energy needs, and consumption, if you use a solar battery. Solar energy batteries have a finite limit of capacity storage space for energy. They can be programmed to switch on at night or at times of day when you need extra energy.

Again, you must be extremely maintenance vigilant when using them, as the risk for electrocution, electric shock and fire are serious.

- **Net metering.** Will you choose to be connected to a utility grid if you install your own solar energy system? Until solar panel battery devices are perfected for their efficiency, safety, and long-term storage capacity, net metering is your best option for using any excess solar generated energy as efficiently as possible. In a way, net metering will allow you to use a utility company's power grid as rudimentary energy storage. One thing is for certain; while you may be able to install your own solar energy system, you will not be able to become net-metering enabled or grid-connected in a DIY fashion. If you change your mind about net-metering, you will have to pay a lot more money to become grid connected and net metering enabled by a professional solar energy system installer after installing your system on your own.

Ask friends, colleagues, neighbors or reputable professional electricians to recommend a setup.

Most solar panel home kits will come with an optional solar permit and contract options as well as optional financing plans. If you decide not to take advantage of such service, you will have to draw up a detailed plan of your system, how you plan to install it and file paperwork for a work permit at your local building inspector office or county or city office concerning construction or building projects on a residential property.

It could take up to two weeks, or much longer considering the DIY nature of the project, for your paperwork to be approved. You cannot begin installation until your permit is approved. A visit from your local building inspector may even have to be planned before, during or after the installation process, depending on where you live.

Be prepared to spend several days, or the equivalent of two long weekends, preparing and installing your DIY solar energy system. A listing of some basic tools that you may require include:

- Solar Panels
- Racks and mounting hardware parts
- Inverter
- Battery

- Charge controller
- Wiring
- Drill
- Ladder

This is not an exhaustive list by any means. The exact tools, equipment, and parts that you will require for your project will depend upon the custom specific

Solar panel modules installed on your roof should be facing the south to receive as much as sunlight energy exposure as possible as the sun transits the sky from east to west. How your panels are placed, where they are placed, and the southerly direction they face will fully maximize how much solar energy they will optimally collect.

You shouldn't just haphazardly place solar panels on your roof without any consideration or pre-planned thought of strategic placement. You may want to spend time taking note of the relative intensity and sky borne transit path of sunlight rays hitting your rooftop.

You will need to develop a multi-step plan for your installation process. Carefully map out and consider how many square feet of space you will ultimately use on your rooftop placing out solar panel modules. You are ultimately drilling holes in your roof to affix solar panels and then threading wiring through your rooftop from the panel modules to an inverter, charge controller and possible battery connection.

Make sure there is no shade or shadow producing obstacles in the way of your planned solar energy system. This would include trees, leaves, and the roof of an adjoining structure and similar obstructions of this nature.

How many solar panels, or modules, that you actually utilize in your installation project will ultimately depend on your projected energy consumption needs. However, since this should be a small-scale DIY project, you will probably only be dealing with several dozen solar panels.

BASIC SOLAR PANEL INSTALLATION GUIDELINES

After you have strategically mapped out the exact rooftop placement points of where you want to place your panel modules, you can start by affixing your rack and mounting framework. Your rooftop rafting will be the guidepost points for strategically placing your racking and mounting outlay. Take your time with this process. Once your racking and mounting outlays are installed, and you have methodically mapped out wiring or drilled holes to thread wiring, you can then install your solar panel modules.

Your solar panel modules will then have to be connected to an inverter. An inverter changes the D/C electrical current generated by a solar panel system into an A/C electrical current. If you are going to connect your solar panels to a battery, you may need a charge controller. A charge controller modulates the flow of electricity flowing into, and out of, the solar battery device.

A building inspector or similar authority will have to ultimate sign off on your solar energy system project before you begin using it. If you plan to connect to a utility grid, you need to file paperwork for that as well too. That will take time. Be prepared to wait through a period of bureaucratic paper-shuffling before you can start using your solar energy system. You could end up waiting a matter of weeks or months.

(Credit: Pixabay)

As mentioned before, a commercially bought solar panel system home kit will cost you anywhere from $5,000 to $20,000 or more. You can buy it online or from a home furnishings or supply department or home supply depot.

Keep in mind that you may pay hundreds or thousands more in additional, unforeseen or extra equipment costs.

Some solar panel systems employ the use of microinverter devices. A microinverter allows for the on-site generation of A/C power, negating the need of an inverter. Also, if one of your solar panels are shaded in any way or malfunction, which is unlikely, the power generation for the whole system can go offline. A microinverter can be installed under each panel and prevent such an occurrence.

Whether a microinverter or standard inverter is acceptable to your needs depends on your home, the state of its wiring and circuit breakers, your energy consumption needs and how much power you plan to generate via your solar energy system.

You may require extra wiring, junction boxes, breakers and numerous additional parts and equipment.

Plan out your project carefully. You should calculate how many years it will take for your DIY solar energy installation to pay back for itself and how much you stand to save against traditional energy costs. Consult the professional advice, counsel, and aid of an electrician and/or solar energy system installer before, during and after installation.

The guidelines that I have provided are just that, basic guidelines to give you a starting point assessment to begin strategically mapping out and plotting your own DIY solar energy system installation project.

Here is a link to an incredible, though truncated, a 5-minute segment of *This Old House* which encapsulates what is involved in a DIY solar panel system roof installation. It also conveys how involved and difficult this process is to accomplish.

I have included this link to give you an idea of what you will be getting yourself into when you commit to a project like this. As well as to convince you to hire or consult the aide of a professional if you are not experienced with solar energy system installations.

This Old House – "How to Install Solar Panels"

https://www.youtube.com/watch?v=subiaaXBoDI

However, you might also save money with solar energy generation if you live rurally, remotely and off the grid. Although saving money on an appreciable level, in a relative long-term time frame, may prove tricky when you take into consideration the maintenance, repair and replacement costs of equipment that has to function long-term remotely, against the elements and off-grid.

I will talk generally about how you can setup and install a commercially produced solar energy system for off-grid use. I will be talking about the general installation of a commercially purchased 5,000kW DIY home solar energy system kit. For an average residential home, 5,000kW may be adequate for 2,500 square feet of space.

For an off-grid, rurally remote DIY project, 5,000kw may be appropriate for 500 to 1,000 square feet. The more space you have, the more kilowatt power you will need. The more kilowatt power you require, the more expensive this process will become.

(**Source:** Pixabay)

An off-grid solar panel energy system will not pay for itself in any appreciable amount of time compared to a connected grid system. As I mentioned before, there are just too many financial considerations to consider for the regular maintenance, repair, and replacement of parts.

Sufficient supplies of diesel for backup generator use will have to be ample, especially if you live remotely or in a self-sustainingly rustic manner. The need for ample and regularly replenished supplies of fossil fuels to maintain an off-grid solar energy system is kind of counterintuitive to the goal of solar energy enabling self-sustainability.

An off-grid solar energy system will have to be upgraded, replaced and repaired as needed and when necessary. You will probably be using backup gas or diesel generators in the event of a breakdown or temporary energy depletion. There will be significant maintenance costs to consider.

Parts will wear down, break down in the elements and have to be replaced. You will have to invest in a storage battery which will probably need to be replaced from time to time. Batteries are expensive and finite in their capacity to store energy, depending on the battery.

Without the benefit of net metering in a grid-connected system, you just won't be using any generated solar power as efficiently as you can. You will have as much access as your battery system will allow you to accommodate. If your battery breaks down, needs repair or replacement, then a lot of time that your system could be used to store excess energy via a connected grid system will be lost.

Remember, current solar energy system battery technology is still evolving. Such batteries are very expensive to buy, maintain, replace. Currently, they are not as energy storage efficient as they can be.

Even if you managed to run an efficient system that would pay off for itself in a decade or two, there are still maintenance, repair, and replacement parts to consider in an off-grid system. Continual maintenance, repair, and replacement costs will increase appreciably off-set any long-term energy savings you manage to incur. This is a system that you will continually have to pay into to maintain its functional viability.

In the end, you will essentially be paying to live self-sustainingly and on your own terms, and not necessarily to have your rurally remote, off-grid solar energy system investment payback for itself decades down the road.

PERFORMING BASIC HOME CHORES UTILIZING SOLAR ENERGY

A solar cooker device

(Credit: Wikimedia)

SOLAR COOKER

It is believed that long ago, human beings began evolving the concept of what we currently call, "cooking," by concentrating the energy of sunlight at strategic times of day onto flat rocks to heat and cook rice and grains.

The harnessed power of the sun was appreciable at the dawn of time. It is a lesson that seems to need to be re-learned in the digital age.

A solar cooker is exactly what it sounds like. It is a rudimentary device that collects and concentrates the heat energy of the sun to cook and heat food or boil liquids. A solar cooker can be a convex mirror or convex-shaped material that is covered in aluminum foil or similar such sunlight energy reflecting material.

A kettle, pot or teapot is then strategically placed near the center of the solar cooker so as to absorb as much heat energy as possible that is being generated. It helps to use cooking pots, pans, kettles and utensils that are colored black to maximize as much solar energy absorption potential as possible.

Solar cookers do not require any fuel for operational purposes. They are pollution free. They are incredibly simple to set up and operate. A solar cooker can achieve temperatures as high as 150 Celsius/302 Fahrenheit and higher, depending on the complexity of setup. You can buy complicated, parabolic mirror shaped solar cooking devices for hundreds of dollars, or more if you wish. You can also just take a cardboard box and cover it in highly sun-reflective material, place it in the direct path of uninterrupted sunlight for multiple hours and place a pot in it as well.

Using the energy of the sun can also help you save energy costs in your solar energy powered house. Indoor cooking can raise heat temperatures indoors significantly, even more so when you take home insulation into account.

In this situation, people compensate by turning on fans or air conditioning, with uses up energy, which you are trying to optimize with a home solar conversion installation. While you may not find yourself cooking every meal you eat with solar cooking, you stand to save considerably on home energy costs in the long run if you cook basic meals on every sunny day that avails itself on a solar cooker.

A solar cooker can take hours to prepare a meal. They should be used to cook and bake non-complex foods with as few ingredients as possible. You can use a solar cooker to thoroughly fry thinly sliced meat proteins or fish, make rice, boil water, bake tortilla bread or pizza, for example. Cooking such relatively simple meals can take hours. A simple solar cooker device is not optimal for use in cooking or baking more complex foods like a pork roast, meatloaf or whole chicken for example. You can cook such things, but you need hours of uninterrupted sunlight to do so.

A solar cooker is only optimal for use on extremely sunny days in areas of high sunlight exposure on non-cloudy days. Your solar cooker must be placed in the direct path of direct sunlight energy to cook as optimally as possible. It can take hours to cook food thoroughly and properly in a solar cooker.

As is normal with cooking activity, you may want to check on your food as it cooks. However, optimal cooking in a solar cooker requires sustained periods of uninterrupted heat-trapping aggregation time. Every time you open a pot of food in a solar cooker, you are allowing the heat energy you are trying to collect to escape

Depending on the simplicity of your solar cooker setup, you may need to anchor or tie down your reflecting material or the solar cooker itself, so it doesn't blow away in the wind on breezy days. Extreme winds and breezy weather may also prematurely cool the food you are preparing or diffuse the solar heat energy you are trying to collect and aggregate in your cookware.

This is a 6-minute YouTube video tutorial by solar cooking enthusiast Barb Ford that offers a basic demonstration on building your own solar cooker.

"Solar Cooker," YouTube video:

https://www.youtube.com/watch?v=DhhXGF8hE20

"Solar Cooker," YouTube video scannable barcode for smart devices:

SOLAR DEHYDRATOR

A solar dehydrator is a device that uses entrapped solar energy to warm air and dehydrate vegetables, fruit or thinly sliced proteins. Solar dehydrating food helps to preserve it for long-term use. The solar dehydrating process also cures or removes water, moisture, and bacteria, from food. If you have ever eaten a sun-dried tomato in an organic or small gourmet restaurant, the chances are that they were probably dried on a rooftop in some rudimentary solar dehydrating device.

There are many kinds of solar dehydrators available. You can buy one for tens or hundreds of dollars. With a little effort, however, you can just build one of your own.

Solar dehydrators require no electricity for use. You can make use of all the food you grow yourself, dehydrate them and save for later use. You will also save hundreds, if not thousands of dollars on saved energy costs related to food refrigeration.

Think of how much money you will save on supermarket trips as well. You will also potentially improve your health, since you will be preserving and eating your own solar dehydrated and preserved food, and eating less unhealthy, processed food.

Your solar dehydrator will have to be placed somewhere sunny for optimal sunlight exposure. It is only practical for use on very sunny days. You also need to take extreme measures to make your solar dehydrator pest and vermin proof.

Air filters and metal screening can be used to keep out vermin, insects, and bugs. Your solar dehydrated food will be left alone for days at a time, so take every precaution to make sure that you are solar dehydrating it for your own consumption and not that of the local vermin.

One kind of solar dehydrator is basically two boxes that are linked together. One box collects and concentrates heat energy, moving that heat energy into another, higher level box, the drying or curing box, with the food to be dehydrated. The flow of solar heated, rising air from the heating box into the dehydration box can be controlled with venting.

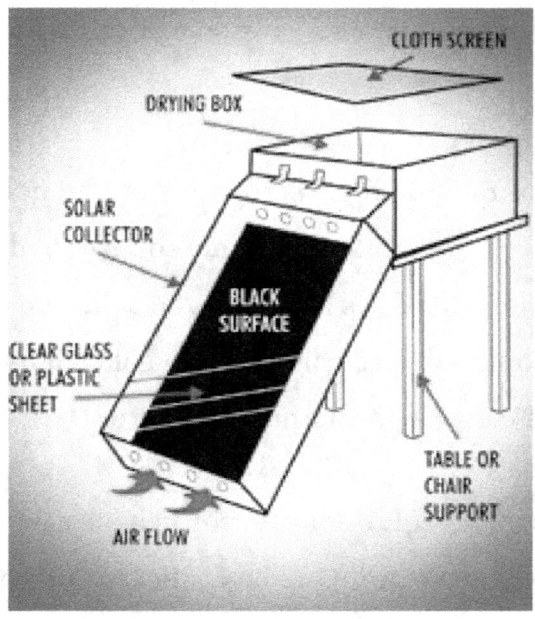

Credit: Off the Grid News

There also single box versions of solar dehydrators that you can build or buy. There are many variations of size and design available for a food dehydrator.

Dehydrating food with a solar dehydrator is an art as well as literal science. It will take you time to get used to using it and determining exactly when food is completely dehydrated. It could take a number of days to completely dehydrate or cure food, so don't rush it.

This is a link to a 2-minute YouTube video produced by Big Brown Farm that demonstrates the solar dehydrating process using a single box method. It is a great starting point to learn more about the solar dehydrating process.

"My Solar Food Dehydrator" 2-Minute YouTube Video

https://www.youtube.com/watch?v=L5vkFPSp3mc

"My Solar Food Dehydrator" 2-Minute YouTube Video Scannable Barcode for Smart Device

HOME SOLAR CONVERSION FAQ

Q: *How long should my home solar energy system last?*

A: As long as you contract your work with a reputable, licensed, insured and recommended solar energy system installer, your solar energy system should last 40 years or longer. That would also depend on the warranty life of the solar panels and materials, the frequency of your sustained and thorough operational maintenance of the equipment and the contractual specifications of the installer to perform regular maintenance and equipment operations checkups.

Home solar energy conversion is a significant investment, like your home. So your solar energy system should be designed to last for as long as you plan to live in your home, which could be more than a decades.

Q: *Exactly how will converting my home for solar energy use affect the overall value of my property?*

A: That will depend on where you live, whether you part of a homeowner's association and whether you intend to ever sell your property, among numerous other factors. If you are part of a homeowner's association, you may have to ask and receive permission before you can begin a home solar energy system installation.

Homeowner association terms, contracts, and guidelines are designed to protect and augment the relative property values of everyone contracted to such. A homeowner's association agreement or contract may have clauses or conditions concerning the maintenance of aesthetic beautification of edifices contracted to the association.

In other words, if you are contracted to a homeowner's association agreement, make sure that there is specific language that will allow you to commence the building of a solar energy installation project. Some people do not like the aesthetics of solar panels on rooftops.

While you may not have a problem with it, other participants of a homeowner's agreement may feel that your solar energy system project, and the visual aesthetic of your solar panels starkly juxtaposed against the natural look of your home's rooftop, may bring down the value of their homes.

(**Credit:** Pixabay)

If you ever plan to sell your property, its value may go up. An already installed solar energy system may be very attractive and appealing to a new buyer who doesn't have to pay for installation.

So, if you attract or look for a buyer who is into solar energy home efficiency, your property value potential may indeed go up. Just be mindful that you have to endure a lot of contractual paperwork to sign over and transfer contractual ownership of the system, and all corresponding solar friendly investment tax credits.

You should also be aware that if you begin a major home construction project, like a home solar energy conversion, then your insurance premium rates might go up.

Insurance actuaries calculate risk and eventuality and data. When you decide to convert your home for solar energy consumption, you may be changing previously accepted home conditions as reflected in an insurance policy prior to construction installation.

Check over your home insurance policy and consult with an agent before you begin a solar energy home conversion project, just to be sure.

Q: *Is there any potential to save even more money on my initial home solar energy system installation setup costs beyond the federally-sponsored, 30% investment tax credit for homeowners*

Yes! December 31, 2021, is the last day for homeowners to take advantage of all available solar friendly investment tax investments. However, there are numerous other local, county, city and municipal tax breaks, incentives and benefits available to take advantage along with the federally-sponsored investment tax incentives.

You have to look for every available solar friend tax incentive that can apply to you. There might be more available for your redemption than you think.

The N.C. Clean Energy Technology Center has developed a searchable database tax incentive seeking homeowners to use called DSIRE or the Database of State Incentives for Renewables & Efficiency®. You can use to input some basic personal information to find every possible solar friend tax incentive that applies to you.

http://www.dsireusa.org/

Q: *Are you sure I can't get paid cash money by my utility company for the excess electricity that I produce via net metering.*

Yes. If you have a home solar energy system, are grid connected and take part in net metering, a utility company will use the meter to track how much excess electricity you generate via solar energy and add to the grid. Your meter will actually run in reverse to track how much extra energy you are adding to the utility company's grid.

Through this process, you are being accountably credited for the amount of energy you are adding to the grid. When you need that extra credited energy, the meter will then run forward, and your credited energy will then start being used.

The utility company will not cut you a check for cash for your excess, solar generated energy. Net metering is a non-cash involved crediting and account system. It just doesn't work like that.

Q: *Is it efficient to be a residential homeowner unconnected to the grid with a solar energy system?*

No, it isn't. For the moment, net metering practices by utility companies are the most optimal way to make the most of the excess electricity that you may generate with your solar energy system. Net metering is federally mandated in most states. This is for the purposes of homeowners being able to efficiently generate as much solar-generated electricity as possible and have later access to it.

You should look at net metering as a utility-grid sized battery for the excess power that your solar power system is bound to generate. Depending on where you live and how much sunlight you receive in your area, you are bound to generate excess energy during certain periods of the day, week, month or year. Via net metering, you can post excess energy to the utility grid which will be credited to you and usable at a later date.

You should remember that a solar energy system is an investment that should pay back for itself in a matter of years or a decade or two. Solar energy battery storage just isn't as efficient as it can be by current standards. You will just never be able to save enough energy for off-peak needs. Modern batteries designed for solar energy storage can store a few hours' worth of energy at best.

If you are not connected to a utility grid as a residential homeowner with a solar energy system, then you are just losing out on long-term savings potential.

Your solar energy system will also probably never pay for itself if you are not connected to the grid. Also, you want to be connected to the grid when your solar energy system doesn't provide enough energy for your calculated needs, instances of inclement weather and for when you need energy at night.

Q: *Are solar energy system batteries worth the purchase?*

A: Current solar power batteries are fine for saving hours' worth of emergency or back up energy.

You have to determine if they are worth the financial cost on or off the grid.

Current solar energy batteries cost thousands of dollars. The starting price for a good one is about $5,000, but you could easily pay double that. Considering how much money you spend as an investment for your solar energy installation project, where is the efficiency is using a $5,000, $6,000 or $7,000 battery that will only save you a few hours' worth of solar-generated electricity?

If you live remotely and off-grid, you will only be able to depend on such batteries for a few hours of charge at a time. It will also be very expensive to repair or replace such batteries.

Also, if you are off-grid and live remotely, you will need more solar panels and more energy to power your property. Imagine paying to purchase multiple solar power batteries at a conservative $5,000 apiece.

If you are not thinking strategically about how you will generate, save and use any solar energy you generate, then you stand to lose out on money and potential energy savings.

Solar energy batteries certainly may have some uses in emergency or back-up use capacities. However, they currently are not efficient enough to justify dedicated and long-term use for storage of any solar energy that you create.

Q: *What will happen to my solar panels in the event of a power outage or blackout? Will they still function if the sun is shining?*

If you are connected to a utility grid, then your solar panels will not function until the grid comes back online. If your solar panels continued to function during a blackout or widespread power outage, then they would continually feed electricity back into the grid. Powerline workers and emergency workers trying to restore power could be electrocuted in such a situation. If you have a battery, and a charge controller or inverter connected to the battery, then you could store power.

Q: *Should I start or finish any home improvement projects before my solar panel system installation?*

Yes. It is a very costly and time-consuming proposition to install, remove and then re-install a solar energy system installation for the sake of unrelated home modeling and construction concerns. For example, if you need to replace your roof, do so before you have any solar panel arrays installed. It just makes sense. It will also save you time, money and headaches.

Q: *How large should my solar energy system be?*

That will depend on a variety of factors. How much energy do you plan to use in your home on a monthly basis? Do you live in an area that experiences a lot of direct sunlight exposure? How big is your home? You have to know the answers to all of these questions, and consult with a solar energy system installer, find out how large a solar energy system you will need.

YOUR FUTURE AS A HOMEOWNER UTILIZING SOLAR ENERGY

I hope I have given you a lot of food for thought when it comes to, "burning the light."

(Credit: upsinverterinfo.com)

Adjusting to the use of a solar energy system will take time to get used to, of course. Solar panel technology is noiseless. It doesn't create pollution or release toxic by-products into the environment. It doesn't involve a myriad of regularly moving parts prone to break down and the need for regular repair and/or replacement.

It may take some getting used to, see your meter run backward in reverse via net metering as you are credited for excess solar energy passed to the grid. Also, maybe, just maybe, if you perfect your skills in the art of solar dehydration, you can get your tomatoes to exhibit that tart and tangy flavor accent.

The utilization of solar energy has the power to change your life completely, if you are willing to embrace it. It would be a good idea to embrace it now and get ahead of the curve on a global scale.

The world is changing. It is changing because the way it uses energy is slowly but surely changing.

The United States government is offering nearly inexhaustible chances for homeowners to take advantage of solar-friendly investment tax credits for a reason. After December 31, 2021, only utility companies and commercial interests will be able to take advantage of permanently set 10% solar friendly investment tax credits. That is going to happen for a reason.

The world is slowly embracing the advent of renewable energy and its eventual replacement of fossil fuels as the planet's prime energy source.

It won't happen tomorrow, next week, next month, next year or even a decade from now. The process has begun, however, and at an appreciable level. Fossil fuel use on the industrial business scale as we know it currently just won't exist in about half a century.

That may feel like a long time, but it isn't. Not really. Not if you live in the same home for 40 or 50 years.

The continued use of fossil fuels is going to get extremely expensive and extremely complicated to deal with in a handful of years.

However, no one has commoditized solar energy. It is free, plentiful and renewable. The harnessing of sunlight energy is truly the last, great, "gold rush," energy resource of this generation of humanity. Probably forever. Get in on the ground floor.

Don't spend the next 40 or 50 years trying to figure out how to make use of fossil fuels and fossil fuel generated energy for your home energy needs in a world that is slowly, but steadily withdrawing from its use.

I know I won't.

Convert your home for solar energy usage. It will be a multi-decade investment like your home, and it will pay back for itself in a handful of years or a decade or two.

Create energy on your own terms and use only the amount of electricity you use for your calculated needs. Save your hard-earned money to do the things you truly want to do in life instead of paying for electricity generated by burning coal.

It won't be easy, but the undertaking of any valuable life investments never are.

Start burning the light.

LAST WORDS

Hopefully, in this book, I was able to give you a good general overview of the basics of solar home conversion.

I wanted to thank you for buying my book; I am neither a professional writer nor an author, but rather a person who always had the passion for alternative energy and saving money. In this book, I wanted to share my knowledge with you, as I know there are many people who share the same passion and drive as I do. So, this book is entirely dedicated to YOU my readers.

Despite my best effort to make this book error free, if you happen to find any errors, I want to ask for your forgiveness ahead of time.

Just remember, my writing skills may not be best, but the knowledge I share here is pure and honest.

If you thought I added some value and shared some valuable information that you can use, please take a minute and post a review on wherever you bought this book from. This will mean the world to me. Thank you so much!!

Lastly, I wanted to thank my wife Jessica and my son Jacob for all their help and support throughout this book, without them, this book would not have been possible.

Thank you once again and be safe.

www.ingramcontent.com/pod-product-compliance
Lightning Source LLC
Chambersburg PA
CBHW071224220526
45468CB00002B/727